My Favorite COLOR WHITE

A Crabtree Roots Book

AMY CULLIFORD

CRABTREE
Publishing Company
www.crabtreebooks.com

School-to-Home Support for Caregivers and Teachers

This book helps children grow by letting them practice reading. Here are a few guiding questions to help the reader with building his or her comprehension skills. Possible answers appear here in red.

Before Reading:
- What do I think this book is about?
    - *I think this book is about the color white.*
    - *I think this book is about things that are white.*
- What do I want to learn about this topic?
    - *I want to learn about animals that are white.*
    - *I want to learn what different shades of white look like.*

During Reading:
- I wonder why...
    - *I wonder why snow is white.*
    - *I wonder why some clouds are white.*
- What have I learned so far?
    - *I have learned that some dogs are white.*
    - *I have learned that snow that falls from the sky is white.*

After Reading:
- What details did I learn about this topic?
    - *I have learned that polar bears are white.*
    - *I have learned that I can find many white things in nature.*
- Read the book again and look for the vocabulary words.
    - *I see the words **polar bear** on page 6 and the word **cloud** on page 8.*

Oct 2021

I see white.

I see white snow.

I see a white
polar bear.

I see a white **cloud**.

I see a white dog.

What do you see
that is white?

Word List

Sight Words

a	is	what
do	see	white
dog	snow	you
I	that	

Words to Know

cloud   **polar bear**

30 Words

I see white.

I see white snow.

I see a white **polar bear**.

I see a white **cloud**.

I see a white dog.

What do you see that is white?

Written by: Amy Culliford
Designed by: Rhea Wallace
Series Development: James Earley
Proofreader: Janine Deschenes
Educational Consultant: Marie Lemke M.Ed.

Photographs:
Shutterstock: Eric Isselee: cover; Marina.Martinez: p. 3; Stoatphoto: p. 4-5; Andrej Prosicky: p. 7, 14; Ratchat: p. 9, 14; PIga Ovcharenko: p. 10; Ranta Images: p. 13

Library and Archives Canada Cataloguing in Publication

Title: White / Amy Culliford.
Names: Culliford, Amy, 1992- author.
Description: Series statement: My favorite color | "A Crabtree roots book".
Identifiers: Canadiana (print) 20210197242 | Canadiana (ebook) 20210197250 | ISBN 9781039600867 (hardcover) | ISBN 9781039600843 (softcover) | ISBN 9781039600850 (HTML) | ISBN 9781039600881 (EPUB) | ISBN 9781039600898 (read-along ebook)
Subjects: LCSH: White—Juvenile literature.
Classification: LCC QC495.5 .C8555 2022 | DDC j535.6—dc23

Library of Congress Cataloging-in-Publication Data

CIP available at Library of Congress

Crabtree Publishing Company

www.crabtreebooks.com   1-800-387-7650

Copyright © 2022 **CRABTREE PUBLISHING COMPANY**

All rights reserved. No part of this publication may be reproduced, stored in a retrieval system or be transmitted in any form or by any means, electronic, mechanical, photocopying, recording, or otherwise, without the prior written permission of Crabtree Publishing Company. In Canada: We acknowledge the financial support of the Government of Canada through the Canada Book Fund for our publishing activities.

Published in the United States
Crabtree Publishing
347 Fifth Avenue, Suite 1402-145
New York, NY, 10016

Published in Canada
Crabtree Publishing
616 Welland Ave.
St. Catharines, Ontario L2M 5V6